Gina goes to the Genetic Counselor

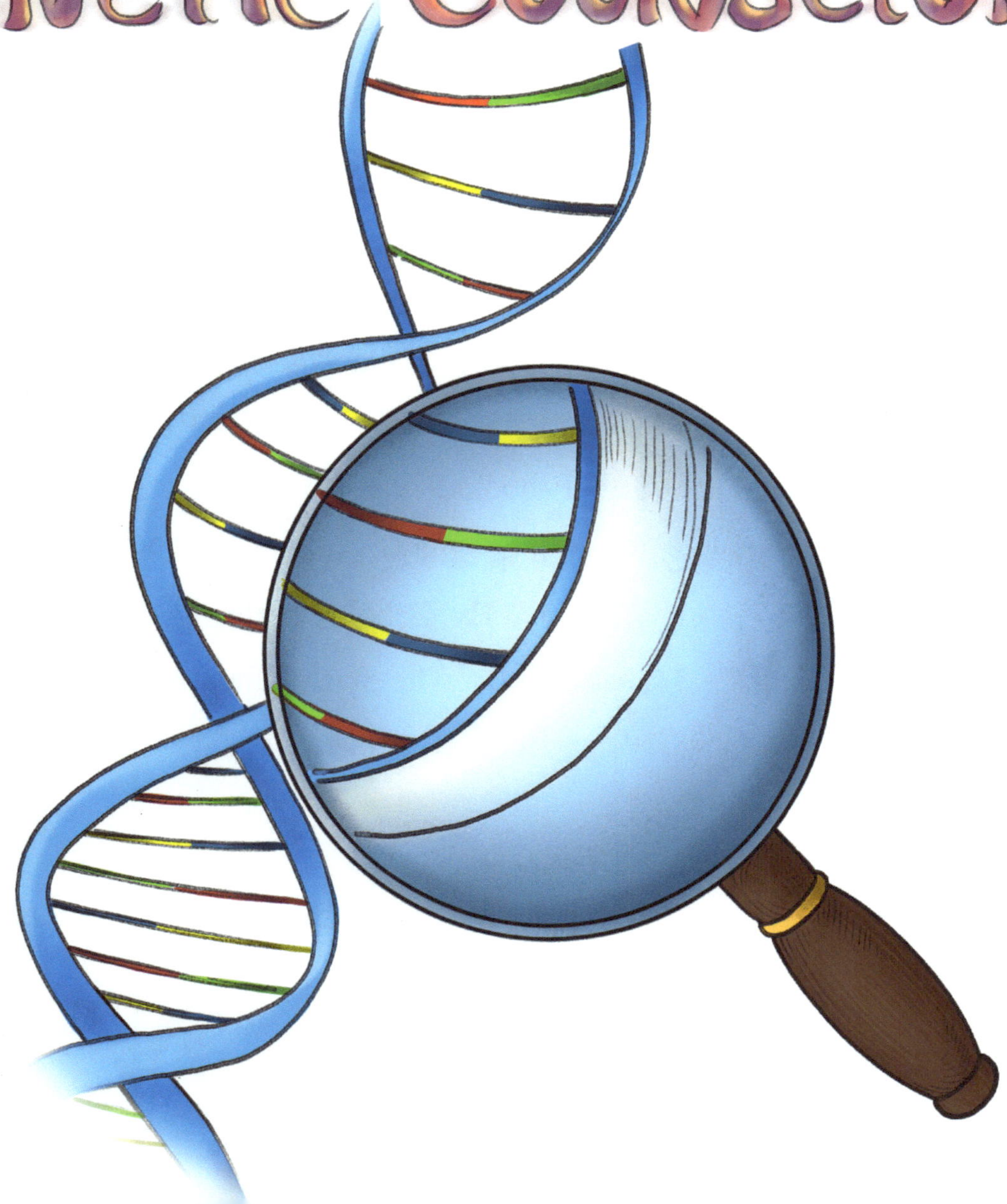

Copyright © 2022 by Sara J. Smith

First Edition

Brooklyn, NY

For permissions contact: Contact@BooksBySara.com

ISBN Paperback: 979-8-9864046-0-8

ISBN E-book: 979-8-9864046-1-5

Library of Congress Control Number: 2022910695

Edited By: Brooke Vitale

Illustrated By: The Stasiuk Family

BooksBySara.com

Dedicated To Families Created Through Adoption

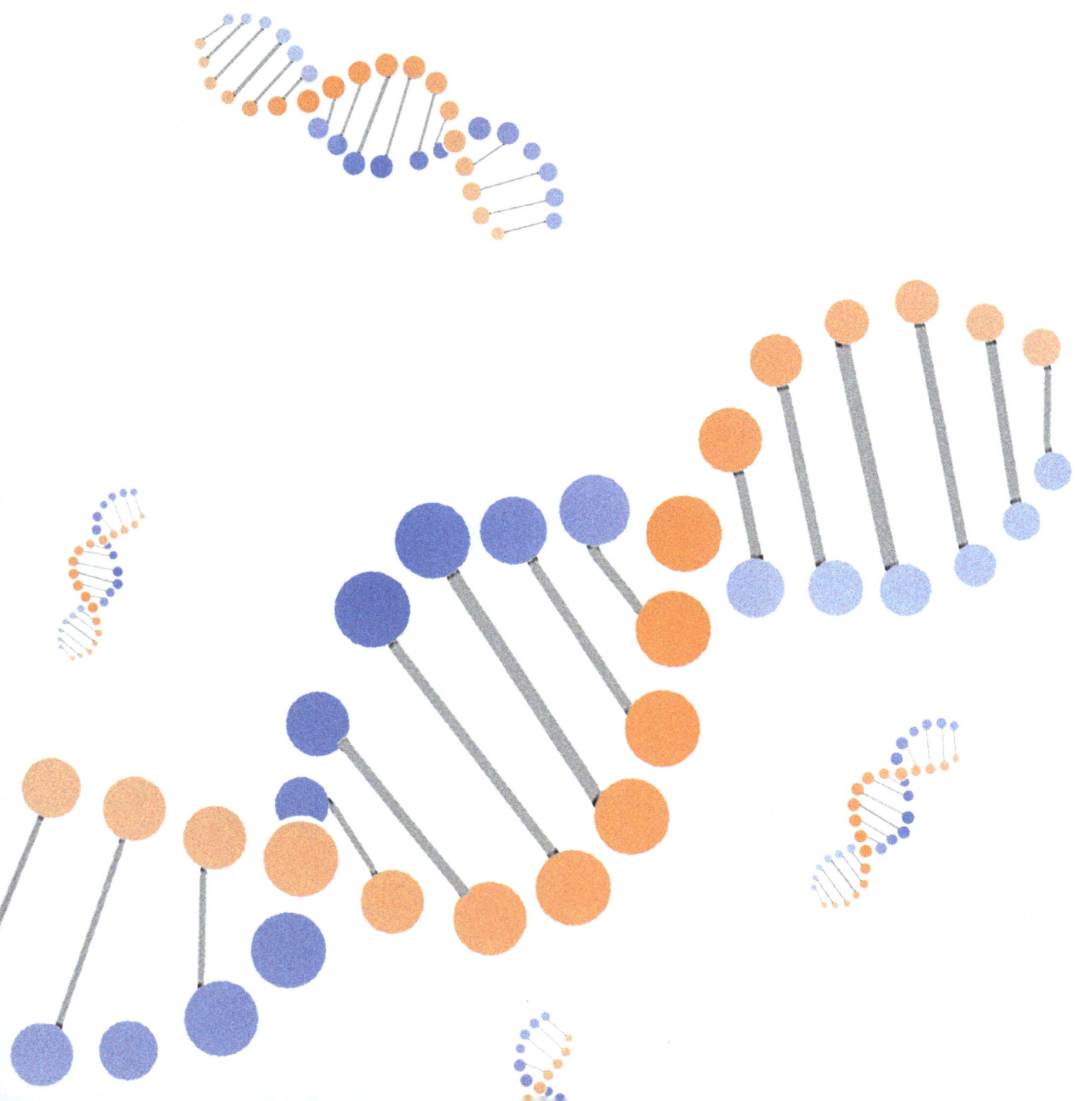

"Congratulations to this season's undefeated soccer team!"

Gina smiled from ear to ear as her parents took photos of her with the championship trophy. Rain or shine, they were always at every game to cheer on Gina.

"Who's that?" Zayn asked suddenly.

Gina wrinkled her nose, confused. "Who's who?"

"Those people taking pictures," Zayn said.

Gina sighed. She couldn't believe that after a full season of soccer, Zayn still hadn't caught on. But the truth was, people were always asking her that question. It wasn't Gina's fault she didn't look like her parents. That's what happens when you're adopted. Even though they looked different, Gina and her parents loved each other to the moon and back!

That night, as Gina's parents tucked her into bed, her dad said, "Remember, we're going to see your doctor in the morning."

Gina nodded, suddenly nervous.

Then, with a last kiss goodnight, her parents turned off the lights.

The next morning, they arrived at the pediatrician's office bright and early.

"What brings you in?" Dr. Lopez asked. "Gina has been dealing with a lot of colds lately," her mom explained. "We aren't sure why she keeps getting sick."

"Gina," Dr. Lopez asked, "how do you feel when you get sick?"

"Really tired," Gina admitted. "Too tired even for soccer."

"And is there a family history of frequent sickness?" Dr. Lopez asked.

Gina's dad shook his head. "We don't know. We adopted Gina when she was a baby and we don't have any family history."

Dr. Lopez sat quietly for a moment. Then she asked, "Have you considered meeting with a genetic counselor?"

"What's a genetic counselor?" Gina asked.

"A genetic counselor is a medical professional who will help you understand how your DNA may affect you," Dr. Lopez explained.

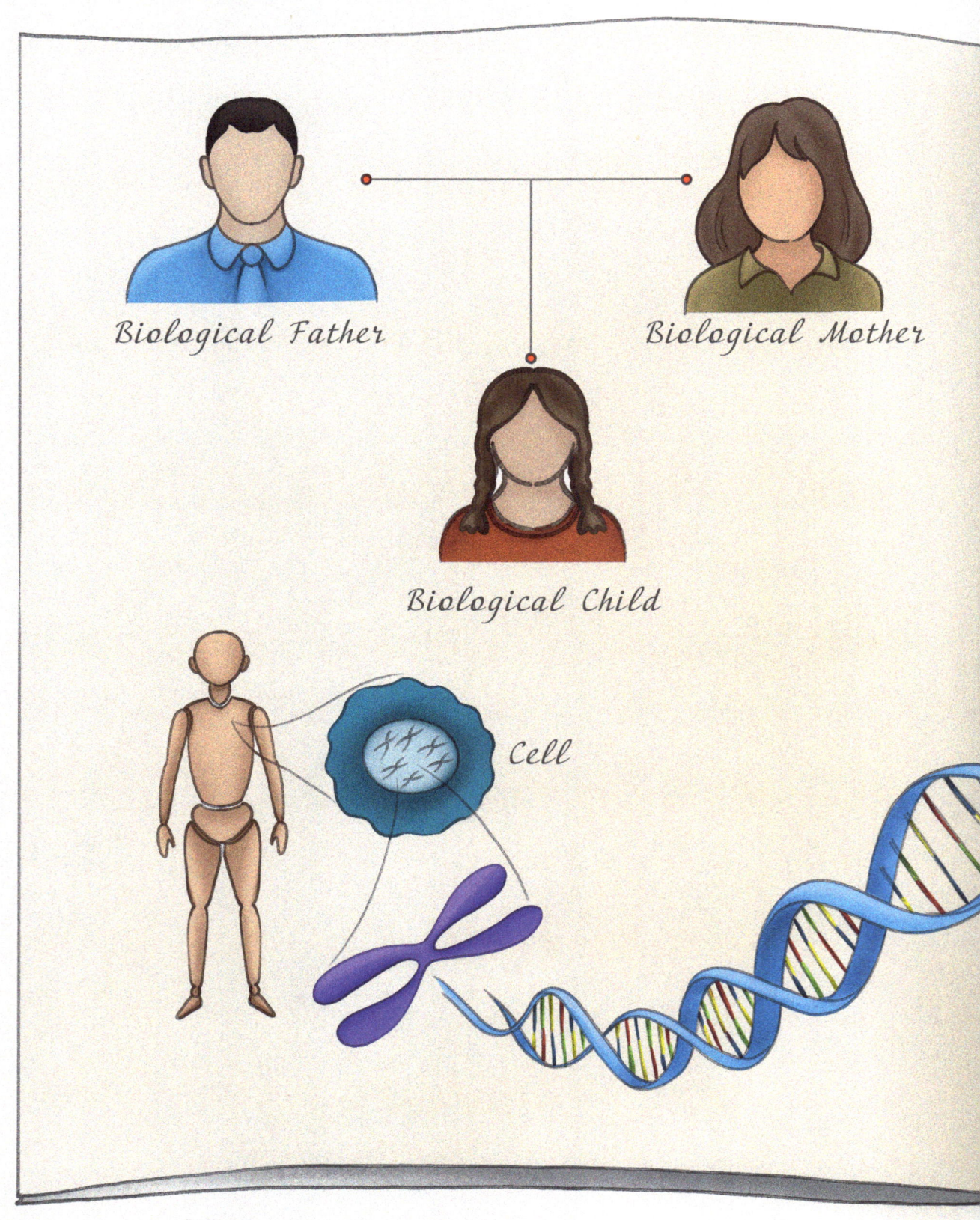

Biological Father

Biological Mother

Biological Child

Cell

"You see, DNA is the body's instruction book. It determines the color of your eyes, the color of your hair, and many other things that make you, you. Half of your DNA is from your biological mother and half is from your biological

DNA

father. Sometimes there are variants—or mutations—in the DNA that can cause health problems. Genetic counselors help people like you and your family by recommending tests and creating personalized care plans to keep you healthy."

That night, Gina lay in bed with her cat, Vinny.

"Sometimes I feel so lonely," Gina admitted. "None of my human friends are adopted like me . . . and you. I really want to find out why I'm always getting sick, but it's kind of scary. What if I'm getting sick because of my DNA? I can't control my DNA. I'm worried about what they might discover."

The next morning, Gina's parents asked if she would like to meet the genetic counselor.

Gina thought about everything she'd shared with Vinny the night before. The thought of meeting the genetic counselor was a bit overwhelming, but she wanted to find out why she was getting sick.

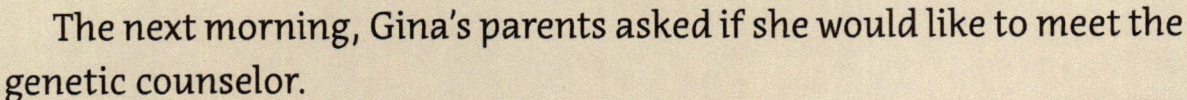

A few days later, Gina found herself sitting across from the genetic counselor, a woman named Kiara.

"Hi, Gina," Kiara said, smiling warmly. "How are you?"

Gina wanted to say, "I'm good," but she didn't want to lie. "I'm nervous," she admitted.

"That's understandable," Kiara said. "What's making you nervous?"

"I don't know anybody who's done this before," Gina said. "What if the information we learn is scary?"

Kiara smiled and listened patiently. "That's understandable because this is a new and different experience. I promise, we won't do any tests or tell you any information without your permission. For now, how about we just talk?"

Gina nodded. Talking sounded good.

For the next half hour, Kiara asked Gina and her parents questions about Gina's health and medical history.

Finally, she said, "There's a test that may be able to help us learn if you have a DNA variant that's causing you to get sick frequently."

"A test?" Gina asked, getting nervous again. "With needles?"

"No needles," Kiara explained. "This test only uses a tube to collect saliva. Would you like to do the test to possibly learn more about what's causing you to get sick?"

"Yes," Gina replied confidently.

With Gina and her parents' permission, Kiara handed Gina a tube and told her to spit in it.

"That's it?" Gina asked.

"That's it. When your results come back in a few weeks, we will go over them together. I will answer all your questions and make you a personalized care plan," Kiara said.

Gina nodded. "Thanks, Kiara."

That night as Gina's parents tucked her in, she realized she felt better than she had in a long time. No matter what the test results showed, Gina and her family felt safe knowing they had a great genetic counselor who would always help them. Kiara would be there for them no matter what. And that felt pretty great!

Glossary

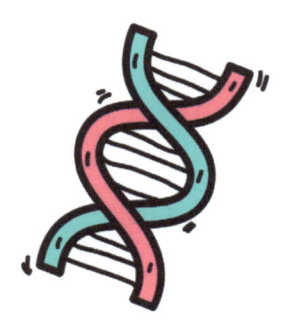

Biological:	genetically related; related by blood
DNA:	abbreviation for deoxyribonucleic acid; molecule that carries genetic information
Genetic counselor:	healthcare professional who helps people understand their genetic background
Genetic test:	medical test to find variants or mutations in an individual's DNA

Resources

- Find A Genetic Counselor (Canada)
  https://www.cagc-accg.ca/

- Find A Genetic Counselor (USA)
  https://findageneticcounselor.nsgc.org/

- Genetic Counseling and Testing
  https://www.cdc.gov/genomics/gtesting/genetic_counseling_testing.htm

- The Adoptee's Guide to DNA Testing by Tamar Weinberg
  ISBN: 978-1440353376

- The DNA Guide for Adoptees by Brianne Kirkpatrick and Shannon Combes-Bennett
  ISBN: 978-1733734301

About The Author

Sara Smith was abandoned at birth in a small, rural Chinese village. She was adopted at fifteen months old from an orphanage in Dianbai, Guangdong, China. She grew up in the United States and went on to graduate magna cum laude from the University of Nebraska at Omaha with a Bachelor of Science in biotechnology with a minor in chemistry while participating in multiple research fellowships. After working in the pharmaceutical industry, Sara decided to change careers and is an aspiring genetic counselor. Her goal is to support families created through adoption during the genetic counseling process.

During her free time she enjoys traveling, practicing Brazilian jiu jitsu, and playing with her cats.

For updates and more, join Sara at BooksBySara.com